Firearms Commerce
in the
United States
2011

United States Department of Justice

Bureau of Alcohol, Tobacco, Firearms

and Explosives

INTRODUCTION

The purpose of this report is to provide an overview of firearms commerce in the United States and the role of the Bureau of Alcohol, Tobacco, Firearms and Explosives (ATF) in regulating firearms commodities. ATF is responsible for enforcing the Gun Control Act (GCA) of 1968, as amended. The GCA regulates the manufacture, importation, distribution, and sale of firearms, and it contains criminal provisions related to the illegal possession, use, or sale of firearms. ATF also administers the National Firearms Act (NFA), which requires the registration of certain weapons, such as machineguns and destructive devices[1], and imposes taxes on the making and transfer of those weapons.

This edition contains information about domestic firearms manufacturing, as well as the importation and exportation of firearms. It also provides an update on certain ATF regulatory initiatives, including our collaboration with the firearms industry, to improve compliance with Federal firearms laws and to prevent the diversion of firearms to illegal markets. Further, it furnishes select data sets regarding domestic firearms manufacturing, and the importation and exportation of firearms.

Part I contains information about domestic firearms manufacturing and the importation and exportation of firearms to provide a better understanding of the firearms market.

Part II provides an overview of recent ATF regulatory program initiatives that focus and improve ATF's ability to conduct firearms inspections, ensure compliance, and foster communication and cooperation with the firearms industry.

Part III describes how ATF's enforcement and regulatory missions work together to prevent the unlawful diversion of firearms into illegal markets.

The appendix to this report presents a series of statistical tables containing the most up-to-date information available about the firearms industry and ATF's regulatory activities. Through these activities, ATF works to ensure compliance with Federal firearms laws and to prevent firearms from being diverted, either knowingly or unknowingly, to persons prohibited from possessing them.

---

[1] Examples of a destructive device include (a) Any explosive, incendiary, or poison gas (1) bomb, (2) grenade, (3) rocket having a propellant charge of more than 4 ounces, (4) missile having an explosive or incendiary charge of more than one-quarter ounce, (5) mine; (b) weapon which may be readily converted to expel a projectile by the action of an explosive or other propellant, and which has any barrel with a bore of more than one-half inch in diameter, etc.

Mission

The Bureau of Alcohol, Tobacco, Firearms and Explosives (ATF) is a law enforcement organization within the U.S. Department of Justice (DOJ). ATF is dedicated to the reduction of violent crime, prevention of terrorism, and protection of our Nation. ATF investigates and prevents crimes that involve the unlawful manufacture, sale, possession, and use of firearms and explosives; acts of arson and bombings; and illegal trafficking of alcohol and tobacco products. ATF regulates the scope of firearms and explosives business activities—from manufacture and/or importation through retail sales; screens and licenses qualified entities that engage in commerce in these commodities; and specifies the form and content of their business records. ATF has established certain standards for the safe storage of explosive materials to which Federal explosive licensees and permittees must adhere.

ATF works closely with government and nongovernment groups in carrying out its regulatory and law enforcement missions. These include industry groups; professional organizations; international, Federal, State, local, and tribal law enforcement authorities; other Federal and public safety agencies; academia; and the communities we serve.

Recent History

On November 25, 2002, President George W. Bush signed into law the Homeland Security Act of 2002. The Homeland Security Act divided the functions of ATF into two agencies.

On January 24, 2003 (60 days after the Homeland Security Act became law), part of the former ATF was transferred from the Department of the Treasury to the Department of Justice and renamed the Bureau of Alcohol, Tobacco, Firearms and Explosives (preserving the acronym "ATF"). ATF was given the authority to administer and enforce the Gun Control Act of 1968 (GCA), the National Firearms Act of 1934 (NFA), and the Arms Export Control Act (AECA) permanent import provisions. ATF was also tasked with overseeing explosives and arson programs and administering the United States Criminal Code provisions relating to alcohol and tobacco smuggling and diversion. For more information visit www.atf.gov.

On the same day, the Alcohol and Tobacco Tax and Trade Bureau (TTB) was created within the Department of the Treasury to handle the regulatory and taxation aspects of the alcohol and tobacco industries. The TTB was entrusted with administering the laws and regulations governing firearms and ammunition excise tax. For more information about TTB activities, visit TTB at www.ttb.gov.

PART I: Firearms Entering Into Commerce

In enforcing the Gun Control Act (GCA) and National Firearms Act (NFA), ATF collects data on the manufacture, importation, and exportation of firearms. This section presents information on current firearms manufacturers and their reported sales volume, along with a review of the necessary procedures for importing and exporting firearms.

Manufacturers' Sales, Exports, and Imports

As of 2011, there are approximately 5,400 licensed firearms manufacturers and 950 licensed importers in the United States. A Federal firearms license is required to engage in the business of manufacturing, importing, or dealing in firearms. These businesses are required by law to maintain records of the production, exportation, importation, acquisition, and disposition of firearms.

Manufacturers' reports to ATF show the number of manufactured firearms disposed of in commerce each calendar year, as well as the number produced for export. In 2009, the most recent calendar year for which information is available, manufacturers disposed of over 2.4 million handguns and more than 3 million rifles and shotguns into commerce. Detailed production information from these manufacturers' reports is contained in the exhibits at the end of this report.

Importation and exportation statistics may also be found in the exhibits. Manufacturers' export volume for calendar year 2009, the most recent year for which complete data is available, reached more than 194,000 firearms. Compared with the 2000 export volume of 188,460 firearms, the increase in firearms exported is significantly less than the increase in firearms imported.

Regulation of Firearms Importation and Exportation

The process of importing and exporting firearms into the United States is regulated by several Federal agencies. ATF administers the import provisions of the GCA, the NFA, and the permanent import provisions of the AECA. This includes the approval or denial of applications to import firearms and ammunition for persons, businesses, and government entities wishing to import such materials into the United States.

ATF also provides technical advice and assistance to the public regarding import requirements applicable to any firearms or ammunition brought into the United States from another country. (See "Technical Assistance to Industry, Law Enforcement, and the Public")

An organization within the Department of State, the Directorate of Defense Trade Controls, regulates the export of firearms other than sporting shotguns. A person wishing to export these firearms must first obtain an export license from the Department of State prior to shipping them. Within the Department of Commerce, the Bureau of Industry and Security regulates the export of sporting shotguns with barrels between 18 and 28 inches in length. A general license from the Bureau of Industry and Security is required to export sporting shotguns.

Statutory Requirements for Importation

The Gun Control Act of 1968 (GCA) generally prohibits the importation of firearms into the United States. However, the GCA creates four categories of firearms that the Attorney General must authorize for importation: These include firearms that are (1) being used for scientific or research purposes, or particular competition or training purposes; (2) unserviceable firearms (other than machineguns) that are designated as curios or museum pieces; (3) firearms that were previously taken out of the United States by the person who is bringing them back; and (4) firearms—other than National Firearms Act and surplus military weapons—that are of a type "generally recognized as particularly suitable for or readily adaptable to sporting purposes" (the "sporting purposes" test). Firearms in this final category comprise the majority of those that are imported into the United States.

Due to a change in the GCA in 1998 that made nonimmigrant aliens prohibited persons, ATF amended its import regulations to require that all nonimmigrant aliens must obtain an import permit to temporarily import firearms for lawful sporting purposes. ATF began processing temporary import applications from nonimmigrant aliens in February 2002. Nonimmigrant aliens must possess a valid hunting license from any state prior to completing the application, or an invitation or registration to a shooting event or trade show, to submit with their application. Prior to this change, these individuals were not required to obtain permits to import firearms temporarily into the United States for sporting purposes, such as for participation in hunting or competitive shooting events. During FY 2010, ATF received 7,228 import applications from nonimmigrant aliens.

Technical Assistance for Industry, Law Enforcement, and the Public

ATF maintains a staff of firearms enforcement officers (FEOs) who provide technical advice and services for manufacturers and importers of firearms, as well as licensed dealers, and the general public. FEOs examine firearms and related products and classify them under the GCA, NFA, and AECA. The FEO also classifies firearms in order to support law enforcement investigations and programs.

PART II: Regulatory Initiatives and Programs

Part II discusses new and ongoing regulatory programs that have improved ATF's ability to ensure firearms industry compliance and to prevent unlawful firearms sales and trafficking.

The Federal Firearms Licensing Center (FFLC)

The Federal Firearms Licensing Center (FFLC) issues Federal firearms licenses. The FFLC processes applications from persons seeking to engage in firearms commerce in accordance with Federal regulations. The FFLC screens all persons who apply for a Federal firearms license to ensure that felons and other prohibited persons do not gain access to firearms.

The Brady Handgun Violence Prevention Act (Brady Act)

The Brady Act requires a background check through the National Instant Criminal Background Check System (NICS), or a State agency, prior to the transfer of a firearm from a FFL to a non-licensee. While the Federal Bureau of Investigation (FBI) or State agencies perform the NICS record checks, ATF analyzes NICS denials, and investigates and enforces Brady Act violations, including the actual or attempted acquisition of a firearm by a prohibited person.

Regulation of Imports, Firearms, and Ammunition

ATF regulates imports of firearms, ammunition, and certain other defense articles through the issuance of import permits. ATF maintains close liaison with the Departments of State and Defense to ensure that import permits do not conflict with the foreign policy and national security interests of the United States.

Licensee Population

ATF is responsible for licensing persons engaged in the business of manufacturing, importing, and dealing in firearms. In addition, ATF issues collector licenses for those who desire to engage in lawful interstate transactions in curio or relic firearms[2].

ATF evaluates information on license applications to determine applicant eligibility for licensing. A review of historical data reveals that the Federal firearms licensee population

---

[2] Firearms which are of special interest to collectors by reason of some quality other than is associated with firearms intended for sporting use or as offensive or defensive weapons. To be recognized as curios or relics, firearms must fall within one of the following categories: (1) Firearms which were manufactured at least 50 years prior to the current date, but not including replicas thereof; (2) Firearms which are certified by the curator of a municipal, State, or Federal museum which exhibits firearms to be curios or relics of museum interest; and (3) Any other firearms which derive a substantial part of their monetary value from the fact that they are novel, rare, bizarre, or because of their association with some historical figure, period, or event.

decreased from a high of more than 286,000 in April 1993, to a low of 102,020 in March 2000, likely due in part to the increase in license fees and requirements to comply with State and local law implemented in 1993 and 1994. Since March 2000, the licensee population has slowly increased to the 2010 level of 118,487 licensees. Trend analysis indicates overall decreases in the number of dealer FFLs and a significant increase in the number of collector FFLs—with approximately 48 percent now licensed as collectors, approximately 45 percent as dealers and pawnbrokers, and 5 percent as manufacturers and importers.

It is also instructive to note the transitory nature of the FFL population. Each year, a significant number of licensees choose to voluntarily cease operations or fail to renew their licenses. For example, during FY 2009, ATF issued over 13,700 new firearms licenses. However, during the same period, the total number of licensees increased by only 2,452.

Firearms Qualification and Compliance Inspection Program

The United States Congress declared that the purpose of the Gun Control Act (GCA) was to provide support to Federal, State, and local law enforcement officials in their fight against crime and violence. Major portions of GCA regulations pertain to licensing and recordkeeping provisions to ensure that individuals who are engaged in the firearms business hold Federal firearms licenses and keep accurate records of firearms acquisition and disposition. Anyone wishing to engage in the activity of importing, manufacturing, or dealing in firearms, or manufacturing or importing ammunition, must obtain a Federal firearms license. Once a license is issued, the licensee has a responsibility to be in full compliance with Federal firearms laws and regulations.

To ensure its mission objectives, ATF conducts face-to-face qualification inspections for all new firearms business applicants. These qualification inspections serve two functions: (1) ensure the applicants understand statutory and regulatory requirements and (2) ensure that only qualified applicants receive a Federal firearms license. ATF conducted almost 7,000 license applicant inspections in FY 2009. During these inspections, ATF Industry Operations Investigators (IOIs) provide extensive education and information about the Federal firearms laws and regulations, and how to stay in compliance with them.

To ensure compliance with recordkeeping requirements, ATF may inspect a licensee once every 12 months. ATF may also inspect a licensee when there is reasonable cause to believe a violation has occurred and a warrant has been issued by a Federal magistrate. Exceptions to the warrant requirement include inspections during the course of a criminal investigation of a person or persons other than the licensee or for determining the disposition of one or more particular firearms in the course of a bona fide criminal investigation.

FFL eZ Check System

FFL eZ Check, which became operational in October 2000, was created to help the firearms industry prevent the fraudulent use of firearms licenses. Prior to a licensee's disposal of a firearm to another licensee, he or she must verify the identity and license status of the person to whom the firearm will be transferred. This is generally accomplished by obtaining a certified copy of the license.

The advent of new computer imaging, scanning, and internet technologies have made it increasingly easier for an unlicensed person to create an authentic looking copy of a license and to fraudulently use that copy to attempt to order firearms from legitimate FFLs.

To help prevent this unlawful activity from occurring, licensees can access ATF's FFL eZ Check at www.atf.gov/applications/fflezcheck. The FFL eZ Check allows an FFL to verify a license prior to transferring a firearm to another licensee. In 2010, FFL eZ Check recorded more than 850,000 license verifications, showing increased usage each year for the tenth consecutive year.

FFL eZ Check also provides a toll-free number (1-877-560-2435) for licensees to call in order to confirm the validity of a particular license. The toll-free number is operational five days a week during standard business hours.

Industry Education

ATF is continually developing and strengthening new working relationships with the firearms industry and consumers through education and outreach. ATF regularly holds informational seminars for licensees to keep them informed about legislative and regulatory changes that affect their businesses. ATF also provides a variety of instructional and informational materials to the industry, including regularly updated reference guides about Federal, State, and local firearms laws and regulations, such as the NFA Handbook and the ATF Guidebook on Importation and Verification of Firearms, Ammunition and Implements of War, licensee newsletters, and a manual for dealers to help them evaluate their vulnerability to thefts. This information is available at www.atf.gov.

To address industry concerns and deal with firearms issues that affect the industry and general public, ATF participates in regularly scheduled meetings with representatives from the National Shooting Sports Foundation (NSSF). ATF also meets with representatives from the National Pawnbrokers Association (NPA), the National Auctioneers Association, and the National Association of Arms Shows (NAAS), in an effort to educate show promoters and attendees about the Federal laws and regulations governing firearms sales at gun shows.

Further, ATF participates in numerous conferences and trade shows sponsored by various firearms industry organizations. The Shooting, Hunting, Outdoor Trade Show and Conference (SHOT Show), sponsored by the National Shooting Sports Foundation, is the largest trade show for all professionals involved with the shooting sports and hunting industries. ATF has been participating in the SHOT Show for over 25 years, regularly hosting an informational booth so that subject-matter experts can be on hand to answer questions and discuss regulatory issues with the 35,000 industry members who attend the show each year.

During the SHOT Show, ATF also attends meetings and conducts seminars to educate industry members about a host of issues such as anti-straw-purchase programs and firearms licensee requirements. Executive-level meetings with the National Firearms Act Trade and Collectors Association (NFATCA) and Firearms and Ammunition Importers Roundtable (FAIR) have become regular features of ATF's SHOT Show participation. In addition, ATF meets collectively with compliance officers from the largest retail firearms dealers on a regular basis.

One example of the positive nature of these activities is the joint ATF-NSSF educational campaign, entitled "Don't Lie for the Other Guy," which is an effort to prevent prohibited individuals from obtaining firearms. This educational campaign is intended to discourage people from illegally purchasing firearms on behalf of others, often for persons who are unable to legally possess firearms. Moreover, it heightens awareness of these illegal sales among licensed dealers. More information about the Don't Lie for the Other Guy campaign can be found at the campaign website www.dontlie.org.

Currently, more than 50,000 Don't Lie Educational Kits have been distributed to FFLs nationwide. FFLs use these kits to train their employees in detecting and deterring illegal straw purchases of firearms. In 2008, the program was expanded to incorporate a consumer awareness component. ATF uses outdoor advertising and public service announcements to convey the important message that buying guns for people who can't or won't buy guns for themselves is a Federal crime punishable by up to 10 years in prison and a fine of up to $250,000. The program alerts the potential straw purchasers of the penalties associated with straw purchases before they enter licensee places of business.

In addition, ATF and NSSF jointly hosted the Don't Lie for the Other Guy educational seminars for licensees at convenient locations around the country. Over 8,000 licensees have attended over 70 of these seminars. The NSSF also prints and distributes the educational kits for firearms retailers. The kits contain posters, pamphlets, and other educational materials including a DVD that depicts five scenarios that straw purchasers use to obtain firearms illegally.

ATF also meets frequently with representatives of the FAIR trade group to discuss issues of mutual interest with respect to the importation of firearms and munitions. Furthermore, ATF attends an annual educational conference for licensed importers and defense contractors along with the Departments of State, Commerce, Treasury, and Homeland Security to discuss issues, concerns, and legislation affecting the import sector of the industry.

ATF continues to work with the firearms industry during times of heightened awareness regarding thefts, unusual purchases, or attempted illegal straw purchases of firearms. Additionally, ATF will continue to notify firearms shippers, when necessary, about the importance of maintaining increased awareness of unusual or suspicious activity. ATF is committed to keeping lines of communication open on issues affecting public safety.

PART III: Integration of Law Enforcement and Regulatory Missions

ATF's Role in Eliminating Illegal Firearms Trafficking

The goal of ATF's illegal firearms trafficking enforcement program is to reduce violent crime and protect national security. ATF investigates and arrests individuals and organizations who illegally supply firearms to prohibited individuals. ATF is the Federal law enforcement organization that regulates the firearms industry. It deters the diversion of firearms from lawful commerce into the illegal market with enforcement strategies and technology. In addition, ATF regulates the firearms industry to promote compliance, to prevent diversion, and to detect those criminals who bring violence to our communities. ATF's illegal firearms trafficking and violent crime strategies provide state and local governments a solution for crime that originates within and outside of their jurisdictions.

Integrated Violence Reduction Strategy (IVRS)

The focus of the Integrated Violence Reduction Strategy (IVRS) is to remove violent offenders from our communities, keep firearms from prohibited persons, eliminate illegal firearms transfers, and prevent firearms violence through community outreach. IVRS builds upon traditional enforcement efforts with the use of State-of-the-art technology, intelligence and information sharing, industry regulation, and community outreach.

The ATF National Tracing Center (NTC)

The ATF National Tracing Center (NTC) is the only firearms tracing facility in the United States. GCA regulations define the required records that allow ATF to trace each firearm from its point of manufacture or importation to the point of its first retail sale. The NTC traces crime guns for Federal, State, local, and international law enforcement agencies to provide investigative leads. By tracing firearms recovered by law enforcement authorities, ATF is able to discern patterns of names, locations, and weapon types. This information provides invaluable leads that help identify persons engaged in the unlawful diversion of firearms into illegal commerce, links suspects to firearms in criminal investigations, identifies potential traffickers, and detects intrastate, interstate, and international patterns in trafficking.

Illegal Firearms Trafficking Information and Intelligence

Because of ATF's unique combination of criminal and regulatory authorities under the GCA, the National Firearms Act (NFA), and the Arms Export Control Act (AECA), it has developed specialized expertise, information, and intelligence resources to more effectively enforce these laws. ATF intelligence research specialists combine ATF proprietary data (e.g.,

Multiple Sales and FFL Out of Business Records) and all source information to identify firearms traffickers, illegal firearms trafficking corridors, and armed violators. ATF intelligence products provide special agents with comprehensive information to detect, investigate, apprehend, and recommend for prosecution those individuals or groups of persons who unlawfully transfer or possess firearms. These tools also assist industry operations investigators (IOIs) in conducting thorough qualification and compliance investigations. ATF's sharing of information and intelligence products contribute to our national security efforts.

Enforcement of the National Firearms Act

By law, ATF regulates National Firearms Act (NFA) weapons such as machineguns, short-barreled rifles, short-barreled shotguns, silencers, certain concealable firearms, and destructive devices. The NFA requires that firearm importers, manufacturers, and makers register NFA weapons. ATF approves or disapproves all NFA transfers and processes all applications and notices to manufacture, transfer, and register NFA items. ATF uses the National Firearms Registration and Transfer Record (NFRTR) to support FFL inspections and criminal investigations. In addition, ATF continually provides technical information to the industry and the public concerning the requirements of the NFA.

Summary

ATF strives to provide the public with the most up-to-date information on the firearms industry, as well as to promote industry collaboration and compliance. ATF will continue to release yearly statistical data relating to the industry, along with information concerning new regulatory initiatives. Any suggestions to improve the usefulness of the data presented in this report may be submitted to:

Bureau of Alcohol, Tobacco, Firearms and Explosives
Office of Enforcement Programs and Services
99 New York Avenue, NE
Washington, DC 20226

Exhibit 1. Firearms Manufactured (1986-2009)

Calendar Year	Pistols	Revolvers	Rifles	Shotguns	Misc. Firearms[1]	Total Firearms
1986	662,973	761,414	970,507	641,482	4,558	3,040,934
1987	964,561	722,512	1,007,661	857,949	6,980	3,559,663
1988	1,101,011	754,744	1,144,707	928,070	35,345	3,963,877
1989	1,404,753	628,573	1,407,400	935,541	42,126	4,418,393
1990	1,371,427	470,495	1,211,664	848,948	57,434	3,959,968
1991	1,378,252	456,966	883,482	828,426	15,980	3,563,106
1992	1,669,537	469,413	1,001,833	1,018,204	16,849	4,175,836
1993	2,093,362	562,292	1,173,694	1,144,940	81,349	5,055,637
1994	2,004,298	586,450	1,316,607	1,254,926	10,936	5,173,217
1995	1,195,284	527,664	1,411,120	1,173,645	8,629	4,316,342
1996	987,528	498,944	1,424,315	925,732	17,920	3,854,439
1997	1,036,077	370,428	1,251,341	915,978	19,680	3,593,504
1998	960,365	324,390	1,535,690	868,639	24,506	3,713,590
1999	995,446	335,784	1,569,685	1,106,995	39,837	4,047,747
2000	962,901	318,960	1,583,042	898,442	30,196	3,793,541
2001	626,836	320,143	1,284,554	679,813	21,309	2,932,655
2002	741,514	347,070	1,515,286	741,325	21,700	3,366,895
2003	811,660	309,364	1,430,324	726,078	30,978	3,308,404
2004	728,511	294,099	1,325,138	731,769	19,508	3,099,025
2005	803,425	274,205	1,431,372	709,313	23,179	3,241,494
2006	1,021,260	385,069	1,496,505	714,618	35,872	3,653,324
2007	1,219,664	391,334	1,610,923	645,231	55,461	3,922,613
2008	1,609,381	431,753	1,734,536	630,710	92,564	4,498,944
2009	1,868,258	547,195	2,248,851	752,699	138,815	5,555,818

Source: ATF's Annual Firearms Manufacturing and Exportation Report (AFMER).

[1] Miscellaneous firearms are any firearms not specifically categorized in any of the firearms categories defined on the ATF Form 5300.11 Annual Firearms Manufacturing and Exportation Report. (Examples of miscellaneous firearms would include pistol grip firearms, starter guns, and firearm frames and receivers.)

The AFMER report excludes production for the U.S. military but includes firearms purchased by domestic law enforcement agencies. The report also includes firearms manufactured for export.

AFMER data is not published until one year after the close of the calendar year reporting period because the proprietary data furnished by filers is protected from immediate disclosure by the Trade Secrets Act. For example, calendar year 2009 data was due to ATF by April 1, 2010, but not published until January 2011.

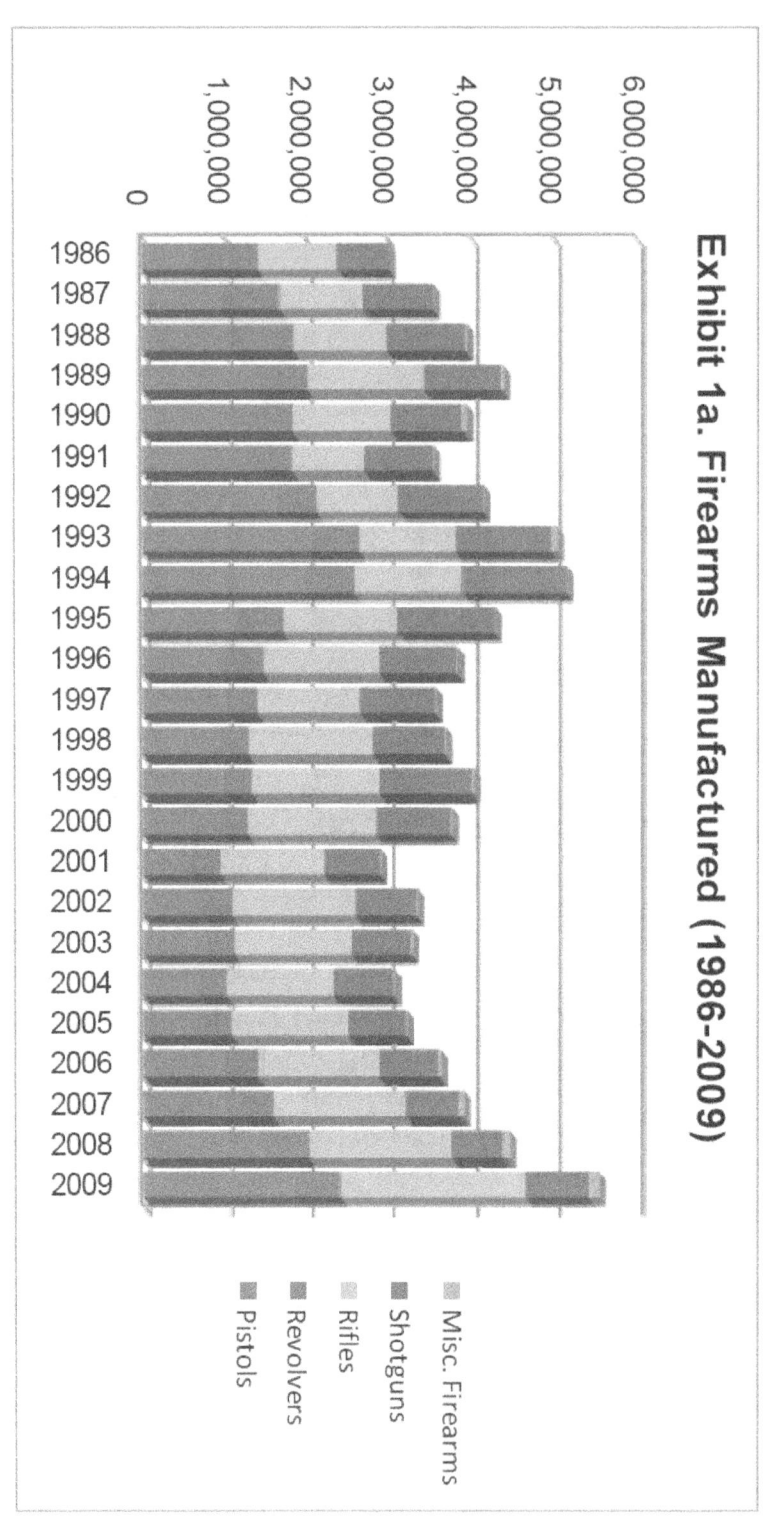

Exhibit 1a. Firearms Manufactured (1986-2009)

Exhibit 2. Firearms Manufacturers' Exports (1986-2009)

Calendar Year	Pistols	Revolvers	Rifles	Shotguns	Misc. Firearms[1]	Total Firearms
1986	16,511	104,571	37,224	58,943	199	217,448
1987	24,941	134,611	42,161	76,337	9,995	288,045
1988	32,570	99,289	53,896	68,699	2,728	257,182
1989	41,970	76,494	73,247	67,559	2,012	261,282
1990	73,398	106,820	71,834	104,250	5,323	361,625
1991	79,275	110,058	91,067	117,801	2,964	401,165
1992	76,824	113,178	90,015	119,127	4,647	403,791
1993	59,234	91,460	94,272	171,475	14,763	431,204
1994	93,959	78,935	81,835	146,524	3,220	404,473
1995	97,969	131,634	90,834	101,301	2,483	424,221
1996	64,126	90,068	74,557	97,191	6,055	331,997
1997	44,182	63,656	76,626	86,263	4,354	275,081
1998	29,537	15,788	65,807	89,699	2,513	203,344
1999	34,663	48,616	65,669	67,342	4,028	220,318
2000	28,636	48,130	49,642	35,087	11,132	172,627
2001	32,151	32,662	50,685	46,174	10,939	172,611
2002	22,555	34,187	60,644	31,897	1,473	150,756
2003	16,340	26,524	62,522	29,537	6,989	141,912
2004	14,959	24,122	62,403	31,025	7,411	139,920
2005	19,196	29,271	92,098	46,129	7,988	194,682
2006	144,779	28,120	102,829	57,771	34,022	367,521
2007	45,053	34,662	80,594	26,949	17,524	204,782
2008	54,030	28,205	104,544	41,186	523	228,488
2009	56,402	32,377	61,072	36,455	8,438	194,744

Source: ATF's Annual Firearms Manufacturing and Exportation Report (AFMER).

[1] Miscellaneous firearms are any firearms not specifically categorized in any of the firearms categories defined on the ATF Form 5300.11 Annual Firearms Manufacturing and Exportation Report. (Examples of miscellaneous firearms would include pistol grip firearms, starter guns, and firearm frames and receivers.)

The AFMER report excludes production for the U.S. military but includes firearms purchased by domestic law enforcement agencies. The report also includes firearms manufactured for export.

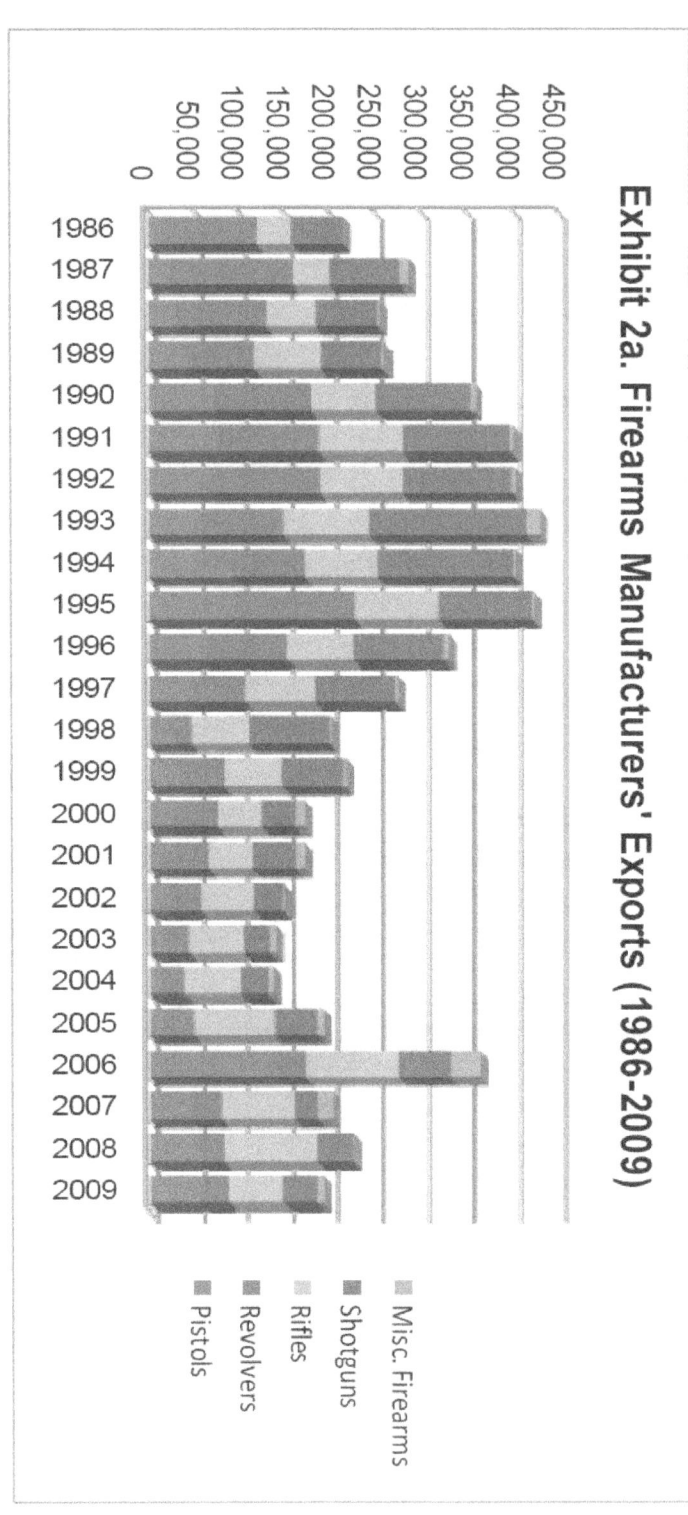

Exhibit 2a. Firearms Manufacturers' Exports (1986-2009)

■ Misc. Firearms
■ Shotguns
■ Rifles
■ Revolvers
■ Pistols

14

Exhibit 3. Firearms Imports (1986-2010)

Calendar Year	Shotguns	Rifles	Handguns	Total Firearms
1986	201,000	269,000	231,000	701,000
1987	307,620	413,780	342,113	1,063,513
1988	372,008	282,640	621,620	1,276,268
1989	274,497	293,152	440,132	1,007,781
1990	191,787	203,505	448,517	843,809
1991	116,141	311,285	293,231	720,657
1992	441,933	1,423,189	981,588	2,846,710
1993	246,114	1,592,522	1,204,685	3,043,321
1994	117,866	847,868	915,168	1,880,902
1995	136,126	261,185	706,093	1,103,404
1996	128,456	262,568	490,554	881,578
1997	106,296	358,937	474,182	939,415
1998	219,387	248,742	531,681	999,810
1999	385,556	198,191	308,052	891,799
2000	331,985	298,894	465,903	1,096,782
2001	428,330	227,608	710,958	1,366,896
2002	379,755	507,637	741,845	1,629,237
2003	407,402	428,837	630,263	1,466,502
2004	507,050	564,953	838,856	1,910,859
2005	546,403	682,100	878,172	2,106,675
2006	606,820	659,393	1,166,309	2,432,522
2007	725,752	631,781	1,386,460	2,743,993
2008	535,960	602,364	1,468,062	2,606,386
2009	558,679	864,010	2,184,417	3,607,106
2010	509,913	547,449	1,782,585	2,839,947

Source: ATF and United States International Trade Commission.

Statistics prior to 1992 are for fiscal years; 1992 is a transition year with five quarters.

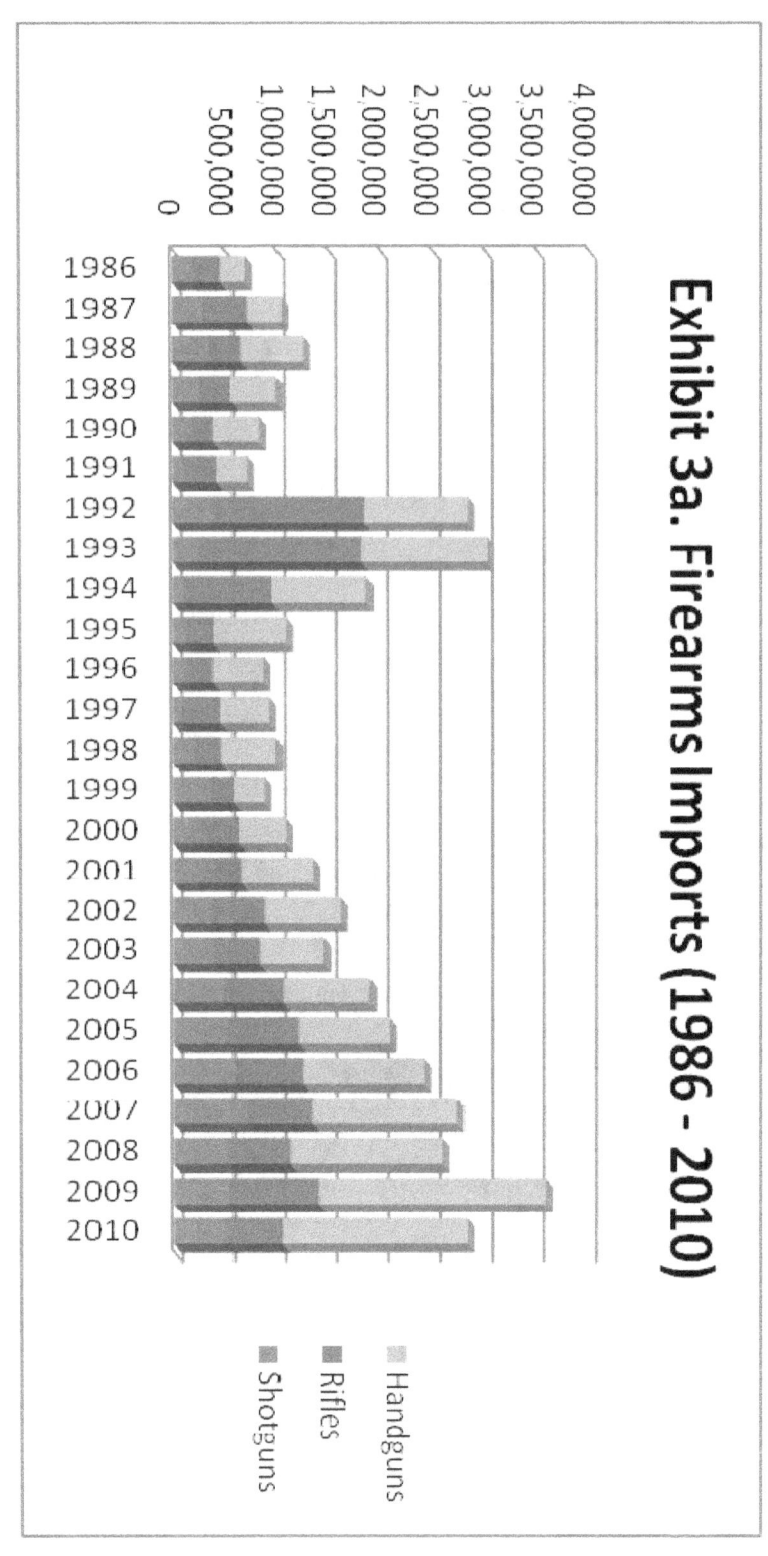

Exhibit 3a. Firearms Imports (1986 – 2010)

16

Exhibit 4. Importation Applications (1986-2010)

Fiscal Year	Licensed Importer	Military	Other	Total
1986	7,728	9,434	2,631	19,793
1987	7,833	8,059	2,130	18,022
1988	7,711	7,680	2,122	17,513
1989	7,950	8,293	2,194	18,437
1990	8,292	8,696	2,260	19,248
1991	8,098	10,973	2,412	21,483
1992	7,960	9,222	2,623	19,805
1993	7,591	6,282	2,585	16,458
1994	6,704	4,570	3,024	14,298
1995	5,267	2,834	2,548	10,649
1996	6,340	2,792	2,395	11,527
1997	8,288	2,069	1,395	11,752
1998	8,767	2,715	1,536	13,019
1999	9,505	2,235	1,036	12,776
2000	7,834	2,885	1,416	12,135
2001	9,639	3,984	1,569	15,192
2002	9,646	6,321	3,199	19,166
2003	8,160	2,264	2,081	12,505
2004	7,539	1,392	1,819	10,750
2005	7,539	1,320	1,746	10,605
2006	8,537	1,180	1,505	11,222
2007	8,004	1,081	1,236	10,321
2008	7,610	718	980	9,308
2009	7,967	504	970	9,441
2010	7,367	823	1,088	9,278

Source: ATF's Firearms and Explosives Import System (FEIS)

Import data excludes temporary permits issued to nonimmigrant aliens.

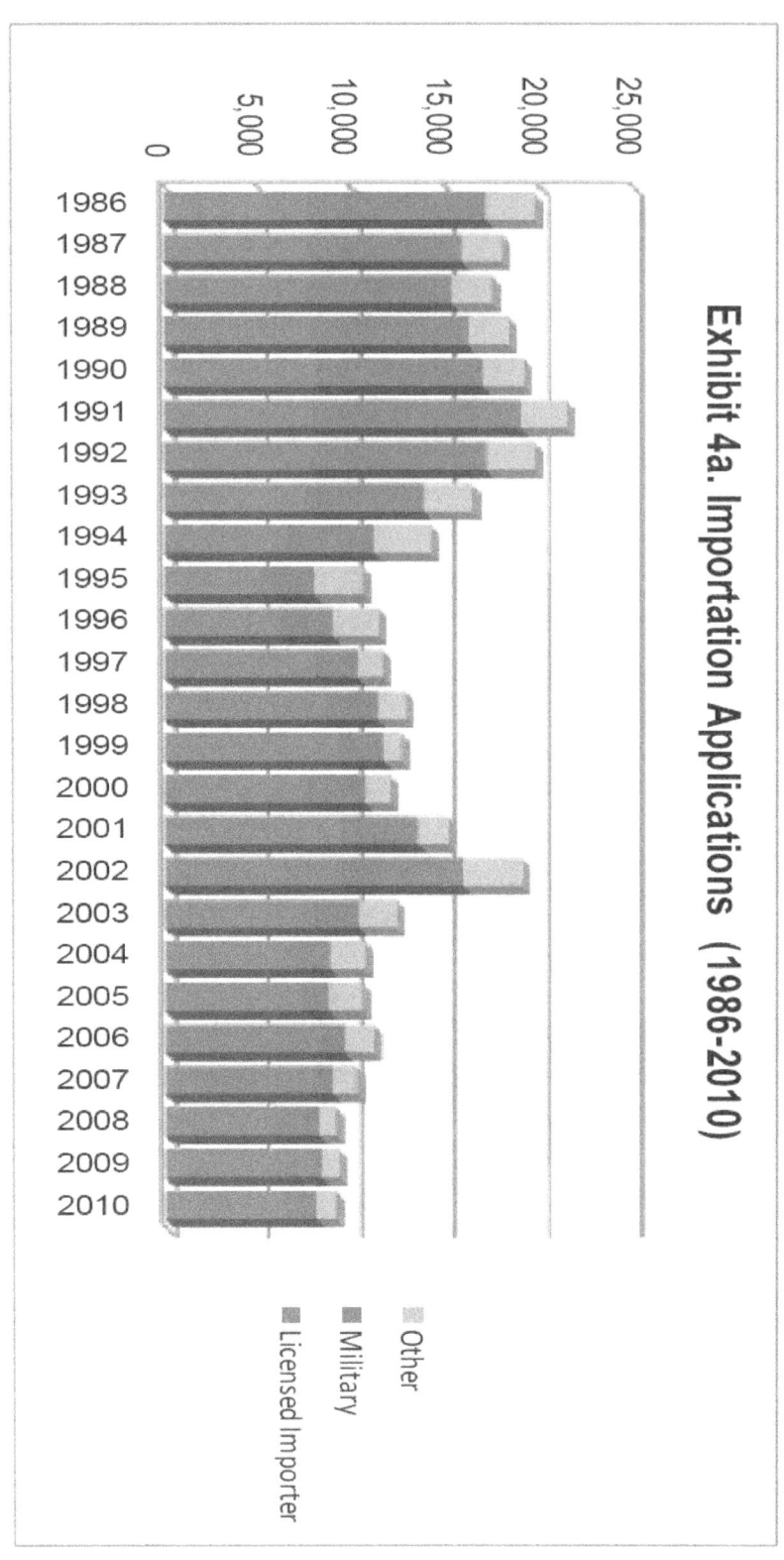

Exhibit 4a. Importation Applications (1986-2010)

Legend: Other, Military, Licensed Importer

Exhibit 5. Firearms Imported into the United States by Country 2010

	Handguns	Rifles	Shotguns	Total Firearms
Brazil	526,011	46,243	169,136	741,390
Austria	431,118	2,759	497	434,374
Italy	129,509	16,393	139,181	285,083
Germany	230,477	33,847	2,364	266,688
Croatia	239,021	0	0	239,021
Turkey	24,443	400	122,721	147,564
Canada	6	154,953	0	154,959
Russia	1,050	90,854	3,708	95,612
Argentina	74,245	0	0	74,245
China[1]	0	300	61,956	62,256
Romania	16,945	35,197	0	52,142
Japan	0	49,946	344	50,290
Philippines	44,626	2,050	1,139	47,815
Serbia	12,455	28,042	0	40,497
Czech Republic	21,140	15,148	34	36,322
Belgium	18,874	16,017	48	34,939
Finland	0	26,464	0	26,464
United Kingdom	387	6,839	6,221	13,447
Spain	989	6,898	1,722	9,609
Ukraine	0	8,800	0	8,800
Portugal	0	4,740	704	5,444
Bulgaria	3,325	0	0	3,325
Poland	3,922	0	70	3,992
Israel	2,645	0	0	2,645
Switzerland	738	1,295	0	2,033
Other[2]	659	264	68	991
Totals	1,782,585	547,449	509,913	2,839,947

Source: United States International Trade Commission

[1] On May 26, 1994, the United States instituted a firearms imports embargo against China. Shotguns, however, are exempt from the embargo.

[2] Imports of fewer than 1,000 per country.

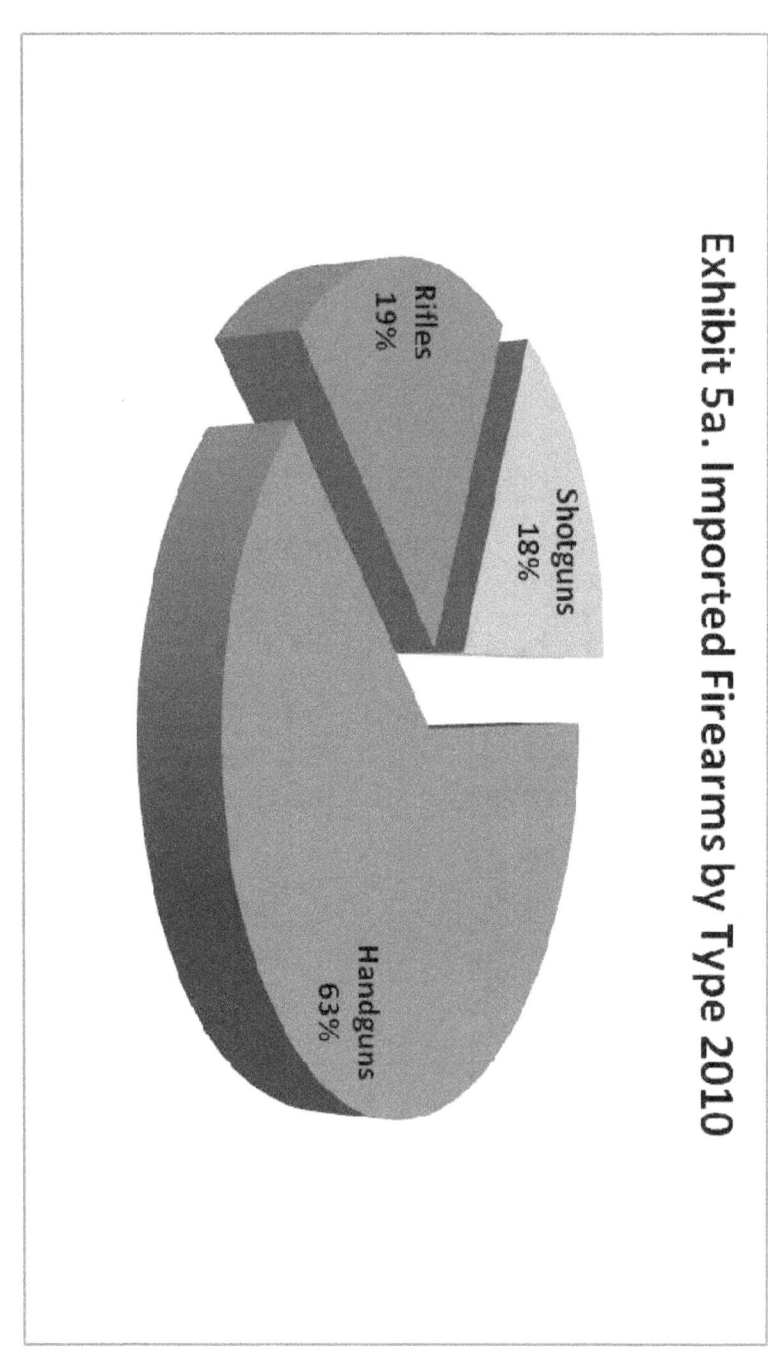

Exhibit 5a. Imported Firearms by Type 2010

Rifles
19%

Shotguns
18%

Handguns
63%

Exhibit 6. National Firearms Act Tax Revenues and Related Activities
(1979-2010)

Fiscal Year [1]	Occupational Tax Paid [2]	Transfer and Making Tax Paid [3]	Enforcement Support [4]	
			Certifications	Records Checks
1979	...	$500,000	3,559	...
1980	...	$716,000	4,377	...
1981	$268,000	$611,000	1,482	3,627
1982	$391,000	$723,000	1,306	2,841
1983	$591,000	$594,000	4,335	...
1984	$596,000	$666,000	1,196	2,771
1985	$606,000	$594,000	921	3,682
1986	$667,000	$1,372,000	690	3,376
1987	$869,000	$1,576,000	575	4,135
1988	$2,095,000	$1,481,000	701	3,738
1989	$1,560,000	$1,527,000	1,196	6,128
1990	$1,442,000	$1,308,000	666	7,981
1991	$1,556,000	$1,210,000	764	7,857
1992	$1,499,000	$1,237,000	1,257	8,582
1993	$1,493,000	$1,264,000	1,024	7,230
1994	$1,444,000	$1,596,000	586	6,283
1995	$1,007,000	$1,311,000	882	5,677
1996	$1,143,000	$1,402,000	529	5,215
1997	$1,284,000	$1,630,000	488	4,395
1998	$1,299,000	$1,969,000	353	3,824
1999	$1,330,000	$2,422,000	345	3,994
2000	$1,399,000	$2,301,000	422	4,690
2001	$1,456,000	$2,800,000	367	2,862
2002	$1,492,000	$1,510,000	503	3,644
2003	$1,758,000	$2,699,000	475	3,749
2004	$1,640,000	$3,052,000	460	3,511
2005	$1,659,000	$2,810,000	471	3,527
2006	$1,709,000	$3,951,000	433	3,349
2007	$1,815,000	$4,890,000	418	3,390
2008	$1,950,000	$5,742,000	304	3,191
2009	$2,125,000	$7,971,000	638	3,175
2010	$2,530,000	$7,183,511	810	3,211

Source: ATF's National Firearms Registration and Transfer Record (NFRTR).

[1] Data from 1997-2000 were based on calendar year data.

[2] Occupational tax revenues for FY 1990-1996 include collections made during the fiscal year for prior tax years.

[3] Importers, manufacturers, or dealers in NFA firearms are subject to a yearly occupational tax.

[4] ATF searches the NFRTR in support of criminal investigations and regulatory inspections in order to determine whether persons are legally in possession of NFA weapons and whether transfers are made lawfully.

Exhibit 7. National Firearms Act Firearms Processed by Form Type (1990-2010)

Calendar Year[1]	Application to Make NFA Firearms[2] (ATF Form 1)	Manufactured and Imported (ATF Form 2)	Application for Tax Exempt Transfer Between Licensees (ATF Form 3)	Application for Taxpaid Transfer (ATF Form 4)	Application for Tax-Exempt Transfer[3] (ATF Form 5)	Exported (ATF Form 9)	Total Firearms Processed[4]
1990	399	66,084	23,149	7,024	54,959	21,725	173,340
1991	524	80,619	19,507	5,395	44,146	40,387	190,578
1992	351	107,313	26,352	6,541	45,390	22,120	208,067
1993	310	70,342	22,071	7,388	60,193	24,041	184,345
1994	1,076	97,665	27,950	7,600	67,580	34,242	236,113
1995	1,226	95,061	18,593	8,263	60,055	31,258	214,456
1996	1,174	103,511	16,931	6,418	72,395	40,439	240,868
1997	855	110,423	18,371	7,873	70,690	36,284	244,496
1998	1,093	141,101	27,921	10,181	93,135	40,221	313,652
1999	1,071	137,373	28,288	11,768	95,554	28,128	302,182
2000	1,334	141,763	23,335	11,246	96,234	28,672	302,584
2001	2,522	145,112	25,745	10,799	101,955	25,759	311,892
2002	1,173	162,321	25,042	10,686	92,986	47,597	339,805
2003	1,003	156,620	21,936	13,501	107,108	43,668	343,836
2004	980	83,483	20,026	14,635	54,675	19,425	193,224
2005	1,902	65,865	26,603	14,606	26,210	20,951	156,137
2006	2,610	188,134	51,290	20,534	100,458	42,175	405,201
2007	3,553	296,267	51,217	22,260	194,794	76,467	644,558
2008	4,583	424,743	71,404	26,917	183,271	206,411	917,329
2009	5,345	371,920	56,947	31,551	201,267	163,951	830,981
2010	5,169	296,375	58,875	33,059	189,449	136,335	719,262

Source: ATF's National Firearms Registration and Transfer Record (NFRTR).

[1] Data from 1990 – 1996 represent fiscal year.

[2] Firearms manufactured by, or on behalf of, the U.S. Government or any department, independent establishment, or agency thereof are exempt from the making tax.

[3] Firearms may be transferred to the U.S. Government or its possessions, to State governments, or to official police organizations without the payment of a transfer tax. Further, transfers of NFA firearms between licensees registered as importers, manufacturers, or dealers who have paid the special occupational tax are l kewise exempt from transfer tax.

[4] Totals do not include ATF Form 5320.20 or ATF Form 10 because these do not relate to commercial transactions.

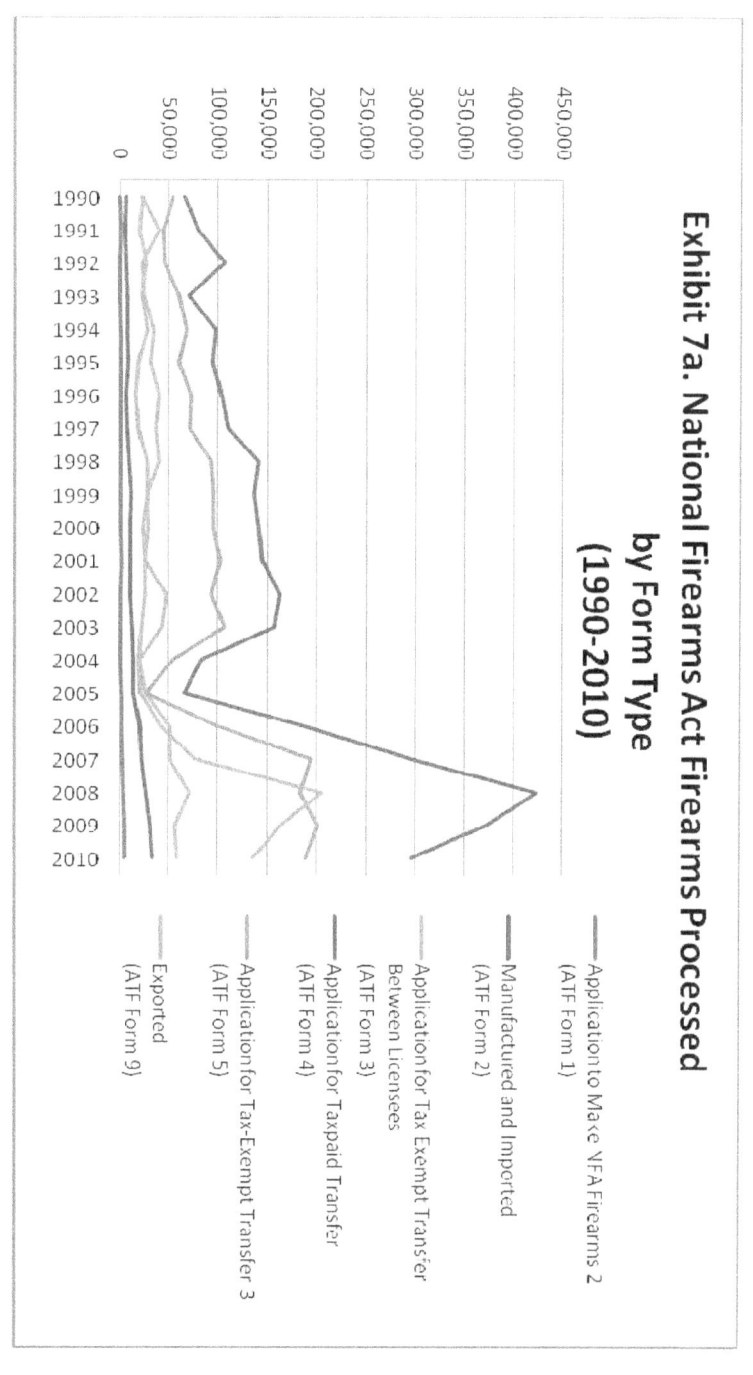

Exhibit 7a. National Firearms Act Firearms Processed by Form Type (1990-2010)

Application to Make NFA Firearms 2 (ATF Form 1)

Manufactured and Imported (ATF Form 2)

Application for Tax Exempt Transfer Between Licensees (ATF Form 3)

Application for Taxpaid Transfer (ATF Form 4)

Application for Tax-Exempt Transfer 3 (ATF Form 5)

Exported (ATF Form 9)

23

Exhibit 8. National Firearms Act Registered Weapons by State (December 2010)

State	Any Other Weapon[1]	Destructive Device[2]	Machinegun[3]	Silencer[4]	Short Barreled Rifle[5]	Short Barreled Shotgun[6]	Total
Alabama	1,085	38,742	14,201	7,195	791	2,013	64,027
Alaska	307	2,840	1,602	1,779	419	1,071	8,018
Arkansas	568	43,567	4,188	5,705	839	977	55,844
Arizona	1,029	63,380	15,633	13,828	4,388	1,747	100,005
California	3,562	195,423	26,237	4,798	2,643	10,103	242,766
Colorado	856	34,346	5,766	4,792	1,675	1,348	48,783
Connecticut	633	8,846	21,943	4,568	878	2,518	39,386
District of Columbia	69	34,059	4,103	160	182	978	39,551
Delaware	32	1,974	515	257	85	441	3,304
Florida	2,906	92,609	27,607	22,517	6,019	5,313	156,971
Georgia	1,687	43,185	20,129	29,004	2,755	9,241	106,001
Hawaii	34	5,019	428	102	50	59	5,692
Iowa	870	11,058	3,055	229	314	882	16,408
Idaho	561	12,432	3,822	8,993	900	379	27,087
Illinois	947	79,858	23,625	1,007	1,100	1,633	108,170
Indiana	1,285	32,239	16,623	11,471	1,886	8,303	71,807
Kansas	670	18,638	2,765	1,284	845	822	25,024
Kentucky	983	20,527	8,706	7,481	1,188	1,615	40,500
Louisiana	511	43,001	6,274	3,302	1,352	1,443	55,883
Massachusetts	817	12,255	6,341	4,769	1,035	844	26,061
Maryland	893	38,657	23,393	5,991	1,408	3,679	74,021
Maine	556	2,342	4,597	1,135	1,370	376	10,376
Michigan	1,038	20,312	8,656	1,322	628	1,094	33,050
Minnesota	2,625	33,944	7,876	632	1,044	995	47,116
Missouri	1,286	23,468	7,660	3,382	1,523	2,088	39,407
Mississippi	387	5,854	3,797	2,416	435	686	13,575
Montana	385	2,652	1,842	1,631	404	338	7,252
North Carolina	801	64,510	10,026	6,862	2,251	2,377	86,827
North Dakota	171	1,433	1,226	1,434	214	149	4,627
Nebraska	703	4,873	2,012	1,792	546	734	10,660
New Hampshire	410	3,203	5,479	2,293	1,121	349	12,855
New Jersey	422	35,733	6,749	850	584	2,267	46,605
New Mexico	269	52,092	3,652	2,451	743	582	59,789
Nevada	674	27,385	6,342	5,575	2,186	742	40,008
New York	1,853	33,689	6,675	892	948	7,243	51,300
Ohio	1,710	67,249	16,254	7,450	2,364	3,589	98,616
Oklahoma	1,065	12,184	6,858	6,615	1,402	1,398	29,522
Oregon	1,446	16,429	6,234	8,391	1,993	1,225	35,718
Pennsylvania	1,891	103,223	17,355	10,059	3,302	11,057	146,887
Rhode Island	43	2,633	572	26	108	112	3,494
South Carolina	650	24,634	6,165	6,096	976	2,233	40,754
South Dakota	339	3,151	1,425	1,336	216	169	6,636
Tennessee	1,436	30,812	11,883	7,765	2,143	5,398	59,437
Texas	6,113	141,891	26,978	36,204	6,958	6,056	224,200
Utah	355	12,629	5,936	4,400	1,576	1,090	25,986
Virginia	2,456	146,071	28,760	13,207	5,807	5,869	202,170
Vermont	217	2,113	1,064	66	112	92	3,664
Washington	1,631	32,863	3,650	5,136	975	728	44,983
Wisconsin	728	22,776	6,067	3,861	1,207	1,087	35,726
West Virginia	420	5,905	2,375	1,784	519	514	11,517
Wyoming	285	95,528	1,634	777	311	369	98,904
Other US Territories	6	286	175	15	11	47	540
Total	52,676	1,864,522	456,930	285,087	74,729	116,462	2,850,406

24

Source: ATF's National Firearms Registration and Transfer Record (NFRTR).

[1] The term "any other weapon" means any weapon or device capable of being concealed on the person from which a shot can be discharged through the energy of an explosive, a pistol or revolver having a barrel with a smooth bore designed or redesigned to fire a fixed shotgun shell, weapons with combination shotgun and rifle barrels 12 inches or more, less than 18 inches in length, from which only a single discharge can be made from either barrel without manual reloading, and shall include any such weapon which may be readily restored to fire. Such term shall not include a pistol or a revolver having a rifled bore, or rifled bores, or weapons designed, made, or intended to be fired from the shoulder and not capable of firing fixed ammunition.

[2] Destructive device generally is defined as (a) Any explosive, incendiary, or poison gas (1) bomb, (2) grenade, (3) rocket having a propellant charge of more than 4 ounces, (4) missile having an explosive or incendiary charge of more than one-quarter ounce, (5) mine, or (6) device similar to any of the devices described in the preceding paragraphs of this definition; (b) any type of weapon (other than a shotgun or a shotgun shell which the Director finds is generally recognized as particularly suitable for sporting purposes) by whatever name known which will, or which may be readily converted to, expel a projectile by the action of an explosive or other propellant, and which has any barrel with a bore of more than one-half inch in diameter; and (c) any combination of parts either designed or intended for use in converting any device into any destructive device described in paragraph (a) or (b) of this section and from which a destructive device may be readily assembled. The term shall not include any device which is neither designed nor redesigned for use as a weapon; any device, although originally designed for use as a weapon, which is redesigned for use as a signaling, pyrotechnic, line throwing, safety, or similar device; surplus ordnance sold, loaned, or given by the Secretary of the Army pursuant to the provisions of section 4684(2), 4685, or 4686 of title 10, United States Code; or any other device which the Director finds is not likely to be used as a weapon, is an antique, or is a rifle which the owner intends to use solely for sporting, recreational, or cultural purposes.

[3] Machinegun is defined as any weapon which shoots, is designed to shoot, or can be readily restored to shoot, automatically more than one shot, without manual reloading, by a single function of the trigger. The term shall also include the frame or receiver of any such weapon, any part designed and intended solely and exclusively, or combination of parts designed and intended, for use in converting a weapon into a machine gun, and any combination of parts from which a machine gun can be assembled if such parts are in the possession or under the control of a person.

[4] Silencer is defined as any device for silencing, muffling, or diminishing the report of a portable firearm, including any combination of parts, designed or redesigned, and intended for the use in assembling or fabricating a firearm silencer or firearm muffler, and any part intended only for use in such assembly or fabrication.

[5] Short-barreled rifle is defined as a rifle having one or more barrels less than 16 inches in length, and any weapon made from a rifle, whether by alteration, modification, or otherwise, if such weapon, as modified, has an overall length of less than 26 inches.

[6] Short-barreled shotgun is defined as a shotgun having one or more barrels less than 18 inches in length, and any weapon made from a shotgun, whether by alteration, modification, or otherwise, if such weapon as modified has an overall length of less than 26 inches.

Exhibit 9. National Firearms Act Special Occupational Taxpayers by State - Tax Year 2010

State	Importers	Manufacturers	Dealer	Total
Alabama	12	27	42	81
Alaska	0	7	24	31
Arizona	13	128	113	254
Arkansas	6	29	29	64
California	11	41	63	115
Colorado	1	32	46	79
Connecticut	8	41	35	84
Delaware	0	1	3	4
District of Columbia	0	0	1	1
Florida	23	127	208	358
Georgia	9	54	80	143
Hawaii	0	0	2	2
Idaho	0	56	30	86
Illinois	9	37	42	88
Indiana	2	38	64	104
Iowa	1	11	11	23
Kansas	3	14	37	54
Kentucky	7	26	45	78
Louisiana	2	22	59	83
Maine	1	15	16	32
Maryland	6	30	40	76
Massachusetts	2	40	22	64
Michigan	6	35	60	101
Minnesota	2	50	15	67
Mississippi	3	14	32	49
Missouri	3	49	63	115
Montana	2	11	21	34
Nebraska	0	8	12	20
Nevada	2	39	40	81
New Hampshire	5	23	35	63
New Jersey	2	5	17	24
New Mexico	3	12	28	43
New York	1	23	13	37
North Carolina	5	63	81	149
North Dakota	0	3	4	7
Ohio	2	70	96	168
Oklahoma	1	33	55	89
Oregon	2	65	53	120
Pennsylvania	8	68	104	180
Rhode Island	1	1	3	5
South Carolina	4	28	18	50
South Dakota	0	6	19	25
Tennessee	4	35	62	101
Texas	9	132	253	394
Utah	4	34	19	57
Vermont	4	12	8	24
Virginia	19	46	88	153
Washington	6	42	24	72
West Virginia	2	15	24	41
Wisconsin	2	45	39	86
Wyoming	1	6	14	21
Total	219	1,749	2,312	4,280

Source: ATF's National Firearms Act Special Occupational Tax Database (NSOT)

Exhibit 10. Federal Firearms Licensees Total (1975-2010)

| Fiscal Year | Dealer | Pawn-broker | Collector | Manufacturer of | | Importer | Destructive Device | | | Total |
				Ammunition	Firearms		Dealer	Manufacturer	Importer	
1975	146,429	2,813	5,211	6,668	364	403	9	23	7	161,927
1976	150,767	2,882	4,036	7,181	397	403	4	19	8	165,697
1977	157,463	2,943	4,446	7,761	408	419	6	28	10	173,484
1978	152,681	3,113	4,629	7,735	422	417	6	35	14	169,052
1979	153,861	3,388	4,975	8,055	459	426	7	33	12	171,216
1980	155,690	3,608	5,481	8,856	496	430	7	40	11	174,619
1981	168,301	4,308	6,490	10,067	540	519	7	44	20	190,296
1982	184,840	5,002	8,602	12,033	675	676	12	54	24	211,918
1983	200,342	5,388	9,859	13,318	788	795	16	71	36	230,613
1984	195,847	5,140	8,643	11,270	710	704	15	74	40	222,443
1985	219,366	6,207	9,599	11,818	778	881	15	85	45	248,794
1986	235,393	6,998	10,639	12,095	843	1,035	16	95	52	267,166
1987	230,888	7,316	11,094	10,613	852	1,084	16	101	58	262,022
1988	239,637	8,261	12,638	10,169	926	1,123	18	112	69	272,953
1989	231,442	8,626	13,536	8,345	922	989	21	110	72	264,063
1990	235,684	9,029	14,287	7,945	978	946	20	117	73	269,079
1991	241,706	9,625	15,143	7,470	1,059	901	17	120	75	276,116
1992	248,155	10,452	15,820	7,412	1,165	894	15	127	77	284,117
1993	246,984	10,958	16,635	6,947	1,256	924	15	128	78	283,925
1994	213,734	10,872	17,690	6,068	1,302	963	12	122	70	250,833
1995	158,240	10,155	16,354	4,459	1,242	842	14	118	71	191,495
1996	105,398	9,974	14,966	3,144	1,327	786	12	117	70	135,794
1997	79,285	9,956	13,512	2,451	1,414	733	13	118	72	107,554
1998	75,619	10,176	14,875	2,374	1,546	741	12	125	68	105,536
1999	71,290	10,035	17,763	2,247	1,639	755	11	127	75	103,942
2000	67,479	9,737	21,100	2,112	1,773	748	12	125	71	103,157
2001	63,845	9,199	25,145	1,950	1,841	730	14	117	72	102,913
2002	59,829	8,770	30,157	1,763	1,941	735	16	126	74	103,411
2003	57,492	8,521	33,406	1,693	2,046	719	16	130	82	104,105
2004	56,103	8,180	37,206	1,625	2,144	720	16	136	84	106,214
2005	53,833	7,809	40,073	1,502	2,272	696	15	145	87	106,432
2006	51,462	7,386	43,650	1,431	2,411	690	17	170	99	107,316
2007	49,221	6,966	47,690	1,399	2,668	686	23	174	106	108,933
2008	48,261	6,687	52,597	1,420	2,959	688	29	189	113	112,943
2009	47,509	6,675	55,046	1,511	3,543	735	34	215	127	115,395
2010	47,664	6,895	56,680	1,759	4,293	768	40	243	145	118,487

Source: ATF Federal Firearms Licensing Center, Federal Licensing System (FLS). Data is based on active firearms licenses and related statistics as of the end of each fiscal year.

Exhibit 11. Federal Firearms Licensees by State 2010

State	FFL Population
Alabama	2,050
Alaska	988
Arizona	2,641
Arkansas	1,744
California	7,177
Colorado	2,295
Connecticut	1,561
Delaware	301
District of Columbia	25
Florida	5,995
Georgia	3,234
Hawaii	256
Idaho	1,265
Illinois	4,103
Indiana	2,541
Iowa	1,844
Kansas	1,699
Kentucky	2,179
Louisiana	1,858
Maine	856
Maryland	2,261
Massachusetts	3,165
Michigan	4,059
Minnesota	2,606
Mississippi	1,340
Missouri	5,500
Montana	1,414
Nebraska	1,061
Nevada	1,160
New Hampshire	969
New Jersey	491
New Mexico	1,005
New York	3,729
North Carolina	3,757
North Dakota	601
Ohio	4,218
Oklahoma	2,104
Oregon	2,317
Pennsylvania	5,446
Rhode Island	432
South Carolina	1,758
South Dakota	678
Tennessee	2,952
Texas	8,383
Utah	1,032
Vermont	526
Virginia	3,629
Washington	2,340
West Virginia	1,347
Wisconsin	2,757
Wyoming	764
Other Territories	74
Total	118,487

Source: ATF, Federal Firearms Licensing Center, Firearms Licensing System. Data is based on active firearms licenses and related statistics as of the end of the fiscal year.

Exhibit 12. Actions on Federal Firearms License Applications
(1975 - 2010)

Fiscal Year	Original Application			
	Processed	Denied[1]	Withdrawn[2]	Abandoned[3]
1975	29,183	150	1,651	...
1976	29,511	209	2,077	...
1977	32,560	216	1,645	...
1978	29,531	151	1,015	414
1979	32,678	124	432	433
1980	36,052	96	601	661
1981	41,798	85	742	329
1982	44,745	52	580	370
1983	49,669	151	916	649
1984	39,321	98	706	833
1985	37,385	103	666	598
1986	42,842	299	698	452
1987	36,835	121	874	458
1988	32,724	30	506	315
1989	34,318	34	561	360
1990	34,336	46	893	404
1991	34,567	37	1,059	685
1992	37,085	57	1,337	611
1993	41,545	343	6,030	1,844
1994	25,393	136	4,480	3,917
1995	7,777	49	1,046	1,180
1996	8,461	58	1,061	629
1997	7,039	24	692	366
1998	7,090	19	621	352
1999	8,581	23	48	298
2000	10,698	6	447	91
2001	11,161	3	403	114
2002	16,100	13	468	175
2003	13,884	30	729	289
2004	12,953	18	572	235
2005	13,326	33	943	300
2006	13,757	35	898	234
2007	14,123	32	953	402
2008	15,434	21	1,030	291
2009	16,105	20	1,415	724
2010	17,025	32	1,468	380

Source: ATF, Federal Firearms Licensing Center, Firearms Licensing System.

[1] Whenever ATF has reason to believe that an applicant is not qualified to receive a license, it may pursue denial of an application. Grounds for denial may include falsification of the application, the prohibited status of the applicant (e.g., felon, drug user, illegal alien, etc.), failure to possess adequate business premises, and others, as prescribed by law.

[2] An application may be withdrawn by an applicant at any time prior to the issuance of a license.

[3] If ATF cannot locate an applicant during an attempted application inspection or cannot obtain required verification data, then the application will be abandoned.

Exhibit 13. Federal Firearms Licensees and Compliance Inspections
(FY 1969 - FY 2010)

Fiscal Year	Licensees	Inspections	Percent Inspected
1969	86,598	47,454	54.7
1970	138,928	21,295	15.3
1971	149,212	32,684	21.9
1972	150,215	31,164	20.7
1973	152,232	16,003	10.5
1974	158,753	15,751	10.0
1975	161,927	10,944	6.7
1976	165,697	15,171	9.1
1977	173,484	19,741	11.3
1978	169,052	22,130	13.1
1979	171,216	14,744	8.6
1980	174,619	11,515	6.5
1981	190,296	11,035	5.7
1982	211,918	1,829	0.8
1983	230,613	2,662	1.1
1984	222,443	8,861	3.9
1985	248,794	9,527	3.8
1986	267,166	8,605	3.2
1987	262,022	8,049	3.1
1988	272,953	9,283	3.4
1989	264,063	7,142	2.7
1990	269,079	8,471	3.1
1991	276,116	8,258	3.0
1992	284,117	16,328	5.7
1993	283,925	22,330	7.9
1994	250,833	20,067	8.0
1995	187,931	13,141	7.0
1996	135,794	10,051	7.4
1997	107,554	5,925	5.5
1998	105,536	5,043	4.8
1999	103,942	9,004	8.7
2000	103,658	3,640	3.5
2001	102,913	3,677	3.6
2002	103,411	5,467	5.2
2003	104,105	5,170	4.9
2004	106,214	4,509	4.2
2005	106,432	5,189	4.9
2006	107,316	7,294	6.8
2007	108,933	10,141	9.3
2008	112,943	11,100	9.8
2009	115,395	11,375	9.9
2010	118,487	10,538	8.9

Source: ATF